Forces and Motion

Think!

What makes an object move?

D0470610

📖 Read the key points. When you finish, check t

Key Points: What is a force?

A **force** is a push or a pull on an object created by its interaction with another object. Applying force to an object can change its **motion**. For example, if you throw a ball to your friend, you are using a pushing force to make it move toward him or her. When your friend catches the ball, he or she stops the ball's motion.

Examples of how forces make objects move.

push

push

pull

Direction is the path an object takes as it moves. If force is applied to an object, it can change its direction. For example, if an object is moving in a straight line and force is applied to it, the object might change direction. This is what happens when you hit a ping pong ball that is coming toward you with your paddle. You are applying a force to change its direction.

✏️ Complete the exercise.

Test your knowledge

Look at where the force is applied to each object and choose the correct direction of motion.

(1) hitting a ball

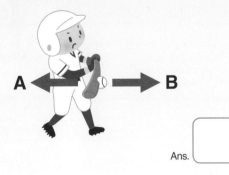

A ←——→ B

Ans. ☐

(2) opening a door

A ←——→ B

Ans. ☐

Chapter 1
Forces and Motion

Think!

Do you think the amount of strength behind a force is important?

📖 Read the key points below. When you finish, check the box.

Check ✓

Key Points: Balanced and Unbalanced Forces

Sometimes more than one force can act on an object. When this happens, the strength and direction of the forces applied to the object will determine if the object moves, and in what direction.

Forces that are equal in strength but opposite in direction are called **balanced forces**. Balanced forces do not cause a change in the motion of an object. When balanced forces act on an object at rest, the object often stays a rest. If balanced forces act on an object in motion, it will continue moving in the same direction and at a constant speed. The key point about balanced forces is that they act with an equal amount of force but in opposite directions.

For example, when two even teams of children pull on a tug-of-war rope with the same amount of force but in opposite directions, the rope will remain evenly spaced between the two teams.

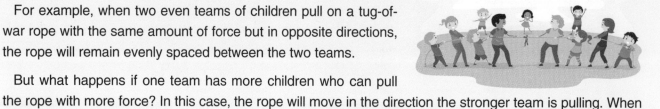

But what happens if one team has more children who can pull the rope with more force? In this case, the rope will move in the direction the stronger team is pulling. When forces are no longer equal, they are considered **unbalanced forces**. Unbalanced forces are not equal. One force is always stronger than the other.

Unbalanced forces change the motion of an object. If an object is at rest and an unbalanced force pushes or pulls the object, it will move. Unbalanced forces can also affect the speed or direction of an object that is already in motion. When two teams of children pull opposite sides of a rope in a tug-of-war the stronger team will always pull the rope with more force, which will cause it to move toward them.

✏️ Complete the exercise.

Test your knowledge

(1) Glenn is walking his dog. The dog starts to pull on the leash, so Glenn pulls back with equal force. This makes them both stop moving foward. Is the force on the leash balanced or unbalanced?

Ans. _____

(2) Sam and Mike are both pushing from opposite sides of a door. The door will not open. Is the force on the door balanced or unbalanced?

Ans. _____

(3) Mila hits a baseball toward center field. Is the force on the ball balanced or unbalanced?

Ans. _____

Chapter 1
Forces and Motion

Think!

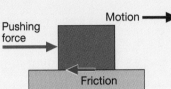

Do you think strength is the only factor that can affect balanced and unbalanced forces?

Check

📖 Read the key points below. When you finish, check the box.

Key Points: Friction

Let's remember that balanced forces acting on an object don't affect the motion of the object. But unbalanced forces do affect an object's motion.

However, this does not mean that once a force is applied to a moving object it will keep moving forever. The object will eventually stop moving due to another force called **friction**.

Friction is the force created when two surfaces rub against each other.

The type of surface an object moves against can affect the amount of friction. **Rough** surfaces create more friction than **smooth** surfaces.

For example, when you kick a ball on a field, it slows down because of the friction created as the ball rubs against the grass. Grass is considered a rough surface, so more friction is created as the ball rolls across it. If the ball was kicked on a smooth surface, like a cement parking lot, it would take longer for it to slow down because there is less friction created between the ball and the cement.

What an object is made of is also important in creating friction. This is why metal ice skates glide on ice, but not cement. This is why you go faster down a slide in athletic shorts than if you slide down in jeans. Jeans are a rougher material, so they create more friction against the slide's smooth surface.

✏️ Complete the exercise.

Test your knowledge

(1) Which surface would create more friction against a rolling ball?

A. ice   B. grass   C. wood   D. tile

Ans. ☐

(2) Which type of footwear would help you slide further across a wood floor?
(Hint: Which material would create the least friction against wood?)

A. socks   B. sneakers   C. ice skates   D. sports cleats

Ans. ☐

Forces and Motion

✏ Use the word box below to fill in the blanks and review key vocabulary.

Review the Key Points

A ⬚ is a push or a pull on an object created by its interaction with another object. Applying force to an object can change its motion. ⬚ is the path an object takes as it moves. If force is applied to an object, it can change its direction.

Forces that are equal in strength but opposite in direction are called ⬚ forces. Balanced forces do not cause a change in the motion of an object. When balanced forces act on an object at rest, the object often stays a rest. If balanced forces act on an object in motion, it will continue moving in the same direction and at a constant speed. The key point about balanced forces is that they act with an equal amount of force but in opposite directions.

When forces are no longer equal, they are considered ⬚ forces. Unbalanced forces are not equal. One force is always stronger than the other.

Unbalanced forces change the motion of an object. If an object is at rest and an unbalanced force pushes or pulls the object, it will move. Unbalanced forces can also affect the speed or direction of an object that is already in motion. ⬚ is the force created when two surfaces rub against each other.

The type of surface an object moves against can affect the amount of friction. Rough surfaces create more friction than smooth surfaces.

> friction / direction / force / balanced / unbalanced

🖩 Complete the exercise.

Math Mission

Different types of surfaces create more or less friction for a moving object. A cardboard ramp was set up to make a toy car roll over different surfaces. Below are the results. Use a ruler to measure the distance the toy car traveled across each type of surface.

(1) A tile floor: _____

Ans. ⬚ cm

(2) A carpet floor: _____

Ans. ⬚ cm

(3) A wool blanket: _____

Ans. ⬚ cm

(4) Based on the results, which surface had the least friction?

Ans. ⬚

Chapter 1
Forces and Motion

Read the mission. Then, answer the following questions to help you with your solution.

The Mission

You are going to take part in a downhill sled race! You want to win and beat last year's winner. The hill has soft and powdery snow and icy patches of snow. Use the knowledge you have learned in this chapter to design a sled that will help you go down the hill fastest and win the race!

Before you design...THINK!

1. Describe the mission in your own words.

2. Brainstorm a solution. Write your notes in the space below.

   Use the following questions to guide your thinking:

   (1) Which type of snow do you think would create the best surface for sliding down the hill?

   (2) What materials can you make your sled out of to help you go faster? Wood, metal, plastic?

Forces and Motion

💡 Read the mission. Then, draw and evaluate your solution.

The Mission

You are going to take part in a downhill sled race! You want to win and beat last year's winner. The hill has soft and powdery snow and icy patches of snow. Use the knowledge you have learned in this chapter to design a sled that will help you go down the hill fastest and win the race!

Design   Draw or write about your solution below.

Evaluate

Did you win the race with the sled you designed? What changes could you make to your sled to make it go faster? Do you think you chose the right type of material for your design?

Simple Machines

Think!

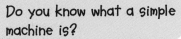

Do you know what a simple machine is?

📖 **Read the key points. When you finish, check the box.**

Check! ✓

Key Points: Forces and Work

Let's say you wanted to move a heavy box up a flight of stairs. You might not be able to produce a large enough force to pick it up and carry it. When we talk about the amount of force necessary to pick up and carry a heavy box, we use the word work. When you move an object, this is called **work**.

Let's look at another example. When you throw a ball, force "changes" the motion of the ball. So, work is done. But, when you push against a wall with just your hands, there is "no change" in the motion of the wall. So, work is not being done. Work is done when a force is applied to an object to make it move. If there is no movement, even if force is applied to an object, the action is not considered work.

This means if we want to carry a heavy box up the stairs, we have to do work. But what if the box is just too heavy for us? This is where **simple machines** can help!

A simple machine is a tool made up of few or no moving parts which can change the strength and direction of a force applied to an object. Remember, a force can be a lift, a push, a pull, or anything that makes an object move.

Simple machines make work, like moving things up stairs, a lot easier. When you use a machine to make work easier, it is called **mechanical advantage**. Mechanical advantage is when you need to apply less force to do the same amount of work. If you use a simple machine to increase or change the direction of the force applied to an object, the work becomes easier.

Let's learn more about the types of simple machines!

✏️ **Complete the exercise.**

Test your knowledge

Choose the word that best completes the sentence.

(1) Work means changing the movement of an object by (energy / force).

Ans. _____

(2) When you pick up a pencil, force changes the motion of the pencil. So work is (done / not done).

Ans. _____

(3) When you push against a wall, there is no change in the motion of the wall. So work is (done / not done).

Ans. _____

Simple Machines

Think!

Can you think of a time you used a simple machine to make your life easier?

📖 **Read the key points below. When you finish, check the box.**

✓ Check

Key Points: Types of Simple Machines

The following are types of simple machines.

A **lever** is a straight bar that can move objects without using a lot of force. All levers have three important parts: a fulcrum (that supports the bar), a load (where the bar touches the object and applies force to the object), and the use of force (the force is applied to the bar). A seesaw is an example of a lever.

A **wheel and axle** is a simple machine that helps objects move. The wheel has a rod or bar called an axle going through it. Similar to a lever, it changes a small amount of force into a larger force. A wheel and axle is commonly found on a car or a wagon. A ferris wheel, like you would ride at a carnival, is an example of a very large wheel and axle!

A **pulley** uses rope and wheels to lift or move objects. Pulleys are used in wells to bring water up from underground. Pulleys are also used to raise flags up flagpoles.

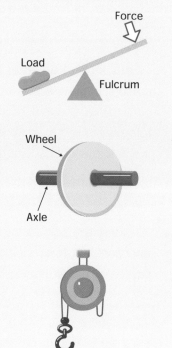

✏️ **Complete the exercise.**

Test your knowledge

Match the type of simple machine to the examples below.

> A: lever   B: wheel and axle   C: pulley

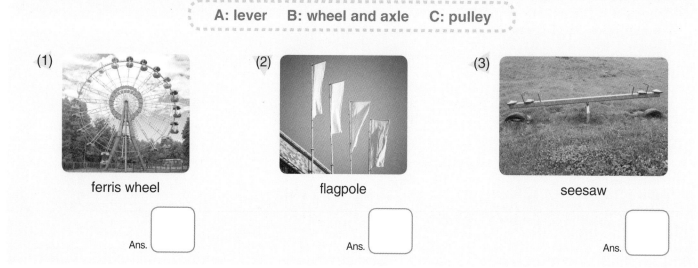

(1) ferris wheel

Ans. ☐

(2) flagpole

Ans. ☐

(3) seesaw

Ans. ☐

Simple Machines

Think!

Can you think of some examples of simple machines in your house?

📖 **Read the key points below. When you finish, check the box.**

Check ✓

Key Points: More Simple Machines

Three other examples of simple machines are the wedge, the inclined plane, and the screw.

A **wedge** has a thick end that narrows to a pointed end which can be driven into an object to split or separate it. Any force applied to the thick end of a wedge is focused into the thin end, and with enough pressure it will split any weaker material it cuts into. An axe is an example of a wedge. Another example of a wedge is a knife you might use to cut your food.

An **inclined plane** is a ramp that is used to take an object from a lower place to a higher place. An inclined plane can also be used to bring an object from a higher place to a lower place. Inclined planes are often used to move large or heavy objects that cannot be picked up easily. Slides and ramps are good examples of inclined planes.

A **screw** has an inclined plane called a thread which is wrapped around an axle. The tip of a screw is often pointed so it can act like a wedge. As you turn the head of a screw, the thread grips into the wood or material you are driving the screw into. This helps hold the material together. Screws are normally used to build objects. Everything from a bookcase to a building can be held together with screws.

Sometimes two or more simple machines can be combined to create a **compound machine**. For example, scissors are a compound machine that uses a lever and wedge. A water faucet uses a wheel and axle, and a screw.

✏️ **Complete the exercise.**

Test your knowledge

Use the word in the box to answer the questions.

> wedge / inclined plane / screws

(1) A ramp for a wheelchair is an example of an [   ].

Ans. [   ]

(2) The axe used to chop down a tree is one example of a [   ].

Ans. [   ]

(3) When you make wooden shelves, you use [   ] to join those planks together.

Ans. [   ]

Simple Machines

✏️ Use the word box below to fill in the blanks and review key vocabulary.

Review the Key Points

[   ] is done when a force is applied to an object to make it move. If there is no movement, even if force is applied to an object, the action is not considered work.

A [   ] machine is a tool made up of few or no moving parts which can change the strength and direction of a force applied to an object. Remember, a force can be a lift, a push, a pull, or anything that makes an object move.

When you use a machine to make work easier, it is called mechanical advantage. If you use a simple machine to increase or change the direction of the force applied to an object, the work becomes easier.

The following are types of simple machines.
- A [   ] is a straight bar that can move objects without using a lot of force.

- A wheel and axle is a simple machine that helps objects move. Similar to a lever, it changes a small amount of force into a larger force.
- A pulley uses rope and wheels to lift or move objects.
- A [   ] has a thick end that narrows to a pointed end which can be driven into an object to split or separate it.
- An inclined plane is a ramp that is used to take an object from a lower place to a higher place.
- A screw has an inclined plane called a thread which is wrapped around an axle. The tip of a screw is often pointed so it can act like a wedge.

Sometimes two or more simple machines can be combined to create a [   ] machine.

> simple / wedge / compound / work / lever

🖩 Complete the exercise.

Math Mission

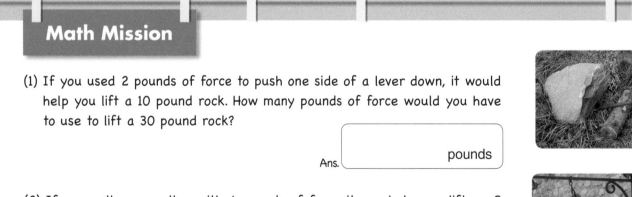

(1) If you used 2 pounds of force to push one side of a lever down, it would help you lift a 10 pound rock. How many pounds of force would you have to use to lift a 30 pound rock?

Ans. [   ] pounds

(2) If you pull on a pulley with 4 pounds of force it can help you lift an 8 pound bucket of water. How many pounds of force would you need to pull with to lift a 40 pound bucket of water?

Ans. [   ] pounds

Simple Machines

💡 **Read the mission. Then, answer the following questions to help you with your solution.**

The Mission

You found a big chest in the attic. However, the chest has no keyhole and cannot be opened by hand. Think about how to open this chest with the help of some simple machines!

Before you design...THINK!

1. Describe the mission in your own words.

2. Brainstorm your solution. Write your notes in the space below. Use the following questions to guide your thinking:

> (1) What simple machines can you use to help you open the chest?
>
> (2) Would you need two or more simple machines?

Simple Machines

💡 Read the mission. Then, draw and evaluate your solution.

The Mission

You found a big chest in the attic. However, the chest has no keyhole and cannot be open by hand. Think about how to open this chest with the help of some simple machines!

Design   Draw or write about your solution below.

Evaluate

What simple machines did you use to move and open the chest? Can you think of a simple machine you could've used instead? Would it have been easier or harder?

Magnets

 Think!

Is there anything you know about magnets? Try talking with your parents and friends.

Check

📖 Read the key points. When you finish, check the box.

Key Points: Magnets

Have you ever used a magnet to hang a piece of paper up on a refrigerator? Or experimented with magnets in your classroom? Let's learn more about how magnets work!

A **magnet** is an object that attracts metal materials and generates a **magnetic field**. A magnetic field is the space around a magnet where the magnetic forces of an object are measureable. Think about how close you can hold the magnet near the refrigerator before it sticks -- that space is its magnetic field! Magnets are attracted to metals that have magnetic properties. These metals include iron, cobalt, steel, and nickel.

Metals without magnetic properties are not attracted to magnets. These metals include aluminum, brass, copper, and lead. Other non-metal materials such as wood, paper, cotton, or plastic are also considered not magnetic.

Strong magnets can work through non-magnetic materials like paper, cloth, aluminum, or plastic. This is why you can use a magnet to hold a piece of paper on a refrigerator. The magnet is strong enough to work through the paper and stick to the metal of the refrigerator door.

✏️ Complete the exercise.

Test your knowledge

Which substance would be attracted to a magnet? Select all correct answers.

A. copper

B. steel

C. wood

D. iron

E. nickel

F. aluminum

G. brass

H. glass

Ans. _____

Magnets

Think!

Can you think of other magnets in your home?

Check ✓

📖 Read the key points. When you finish, check the box.

Key Points: Magnet Properties

All magnets have a **north pole** (N side) and a **south pole** (S side). The sides of a magnet are named this way because one end of a suspended magnet always points northward while the other points southward. Magnet poles always align with the Earth's natural magnetic poles.

Opposite poles or sides attract while same poles repel. When two magnets or a magnet and a metal object attract, they move toward each other. When two magnets repel, they are forced away from each other.

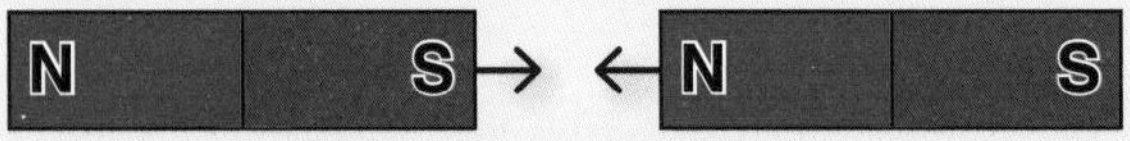

Opposite poles **attract**

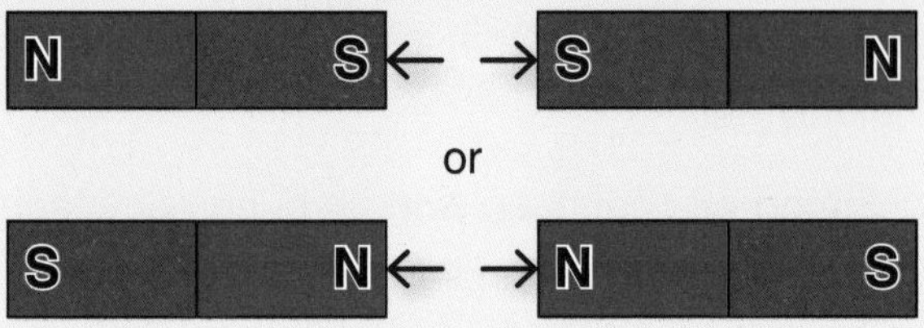

or

Same poles **repel**

✏️ Complete the exercise.

Test your knowledge

Determine if the magnets below will attract or repel each other.

(1)  N   S ➡ ? ⬅ S   N   Ans. _____

(2)  S   N ➡ ? ⬅ S   N   Ans. _____

(3)  N   S ➡ ? ⬅ N   S   Ans. _____

(4)  S   N ➡ ? ⬅ N   S   Ans. _____

Chapter 3
Magnets

Think!

What are some ways people use magnets?

📖 **Read the key points. When you finish, check the box.**

Check ✓

Key Points: Uses for Magnets

Magnets are important tools for scientists and engineers. They can help make tasks easier or help solve a problem. Engineers have even developed ways to store information on computers using magnetic memory technology!

Believe it or not, your home contains many magnets you use every day!

Refrigerator magnets hold papers and other small items to a metal refrigerator door. Other tools like a compass use a magnetic needle to show which way is north. The dark magnetic strip on the back of a credit card stores data in much the same way as a computer's hard drive does. Vacuum cleaners, blenders and washing machines all have electric motors that work by magnetic principles. You'll find magnets in cell phones, doorbells, shower curtain weights, and even some of your toys.

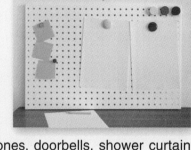

✏️ **Complete the exercise.**

Test your knowledge

Choose all the items that use magnets from the following items found in daily life.

A. vacuum cleaner

B. computer

C. credit card

D. washing machine

E. refrigerator

F. cell phone

G. blender

H. compass

Ans.

Magnets

✏ Use the word box below to fill in the blanks and review key vocabulary.

Review the Key Points

A magnet is an object that attracts metal materials and generates a [_____]. A magnetic field is the space around a magnet where the magnetic forces of an object are measureable. Think about how close you can hold the magnet near the refrigerator before it sticks -- that space is its magnetic field! Magnets are attracted to [_____] that have magnetic properties. These metals include iron, cobalt, steel, and nickel.

Metals without magnetic properties are not attracted to magnets. These metals include aluminum, brass, copper, and lead. Other non-metal materials such as wood, paper, cotton, or plastic are also considered not magnetic.

All magnets have a [_____] (N side) and a south pole (S side).

Opposite poles or sides attract while same poles repel. When two magnets or a magnet and a metal object attract, they move toward each other. When two magnets repel, they are forced away from each other.

[_____] are important tools for scientists and engineers. They can help make tasks easier or help solve a problem. Engineers have even developed ways to store information on computers using magnetic memory technology!

Believe it or not, your home contains many magnets you use every day!

> magnets / magnetic field / north pole / metals

🖩 Complete the exercise.

Math Mission

(1) Is it possible to hang a 0.4 lb wallet in a 1.5 lb handbag on a magnetic hook that can withstand up to 2 lbs?

Ans. | Yes / No |

(2) Is it possible to hang a 0.9 lb folding umbrella and a 0.4 lb key case together on a magnetic hook that can withstand a weight of 1.2 lb?

Ans. | Yes / No |

Magnets

💡 Read the mission. Then, answer the following questions to help you with your solution.

The Mission

You accidentally spilled a box of sewing pins on the rug! It will take too long to pick them up one-by-one, and you might prick your fingers! Design a device using a magnet to help you pick up all the pins lost in the rug.

Before you design...THINK!

1. Describe the mission in your own words.

2. Brainstorm your solution. Write your notes in the space below.

Use the following questions to guide your thinking:

(1) What type of magnet would you use?

(2) How could you pick up the most pins in the least amount of time?

Magnets

💡 **Read the mission. Then, draw and evaluate your solution.**

The Mission

You accidentally spilled a box of sewing pins on the rug! It will take too long to pick them up one-by-one, and you might prick your fingers! Design a device using a magnet to help you pick up all the pins lost in the rug.

Design

Draw or write about your solution below.

Evaluate

Did your tool pick up all the pins? How could you improve your tool? Can you add to it or take away a piece from it to make it work better?

Forms of Energy

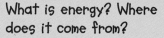

Think!

What is energy? Where does it come from?

📖 **Read the key points. When you finish, check the box.**

Check

Key Points: What is Energy?

In Chapter 2, you read about work and how a force is necessary to make work happen. But, did you know that in order to produce a force you need energy? So, what is energy and where does energy come from?

Energy is the ability to do work. Energy makes things change and move. It's everywhere around us and takes many forms. Here are some examples:

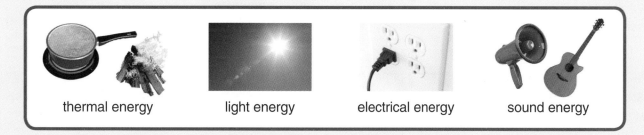

thermal energy   light energy   electrical energy   sound energy

Energy cannot be created from nothing, and it cannot be destroyed. All the energy available in the world already exists! In this way, the energy available to do work always comes from somewhere else, and that energy comes from another form of energy, and then from a previous form of energy before that. For example, when you eat an apple your body takes the chemical energy from the apple and changes it into mechanical energy to help you make your bed or carry out an action. We use the natural properties of energy to turn one form of energy into another, useful form of energy every day. This change from one form of energy to another is called **energy conversion**.

✏️ **Complete the exercise.**

Test your knowledge

Match the form of energy with each example.

(1) light energy   ●

(2) sound energy   ●

(3) thermal energy   ●

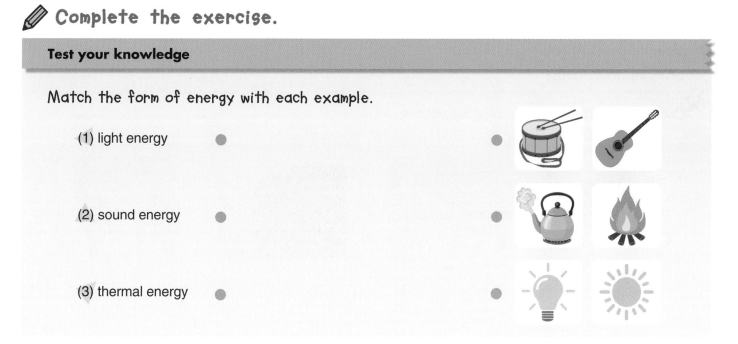

Chapter 4
Forms of Energy

Think!

Can you think of examples of energy you use in your everyday life?

Check

Read the key points. When you finish, check the box.

Key Points: Heat Energy

Let's explore different types of energy.

Heat energy is a form of energy that is produced by the flow of energy from a warm object to a cooler object. Heat energy or thermal energy can be transferred from one object to another. Energy will flow from the warmer object to the cooler object until both objects reach the same temperature. The transfer, or flow, of energy is caused by the temperature difference between the two objects. **Temperature** is the measure of how hot or cold a substance is.

When energy moves from a warm object to a cooler object it heats the cooler object. The amount of energy used to change the temperature of different objects varies based on what substance they are made of. For example, a concrete sidewalk heats up faster in the summer than the grass next to it. This is because the solid concrete will absorb the sun's rays more quickly than the grass.

Heat also travels between different types of substances. For example, you might use a pot to make hot water on a kitchen stove. When the heat is transferred from the stove to the metal pot, the temperature of the water in it also rises. Once the pot is hot, the water in it will also be hot.

Did you know you can prevent heat from being transferred to other substances? A material that prevents heat from being transferred is called an **insulator**. Cotton, wool, and air are good insulators. Metal is a poor insulator because it is a **conductor** which holds heat easily. So, if you need to pick up a pot of boiling water, please use an insulator like a oven glove or a pot holder, which are typically made of cotton and will protect your hand from the heat.

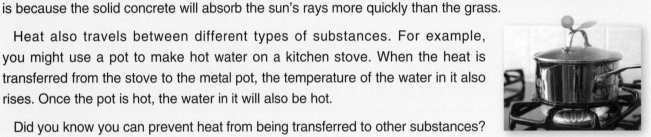
Complete the exercise.

Test your knowledge

Choose all materials that are insulators.

A. wool

B. air

C. metal

D. cotton

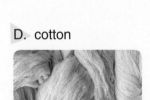

Ans.

Forms of Energy

Think!

Did you know light was a form of energy?

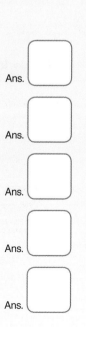

Check

📖 Read the key points. When you finish, check the box.

Key Points: Light Energy

Light is another type of energy. Light energy, or radiant energy, is the only type of energy we can see with our eyes. A natural example is the light that comes from the sun and the stars. Although it can be scary, a flash of lightning is also light energy. There are many man-made light sources as well, such as light bulbs and candles. Light energy allows us to fill dark spaces with light, so we can see what is around us.

Light travels in a straight line from its source until it hits an object or surface it cannot penetrate, or pass through. For example, light can pass through glass windows, but not concrete walls. If you enter a room with concrete walls that has no windows or lights, you will not be able to see anything because no light can enter the room. Any material that doesn't allow light to pass through, like concrete, is called **opaque**. Wood, brick, cardboard, and tin are examples of opaque materials.

Sometimes light can hit an object and bounce off its surface instead of passing through it. This phenomenon is called **reflection**. Reflection causes light to change direction or bend. You can see reflection at work if you use a mirror to reflect light from the sun into a dark room or onto another surface. The light shining on the mirror will bounce off its surface and shine wherever you direct it.

✏️ Complete the exercise.

Test your knowledge

Answer T for true or F for false.

(1) Light energy is the only type of energy we can see with our eyes.

Ans. ☐

(2) Light travels in a curved line from its source until it hits an object or surface it cannot penetrate, or pass through.

Ans. ☐

(3) Air is an example of an opaque material.

Ans. ☐

(4) Cardboard is an example of an opaque material.

Ans. ☐

(5) A mirror is an object that can reflect light.

Ans. ☐

Chapter 4
Forms of Energy

✎ **Use the word box below to fill in the blanks and review key vocabulary.**

Review the Key Points

Energy cannot be created from nothing, and it cannot be destroyed. All the energy available in the world already exists! In this way, the energy available to do work always comes from somewhere else. We use the natural properties of energy to turn one form of energy into another useful form of energy every day. This change from one form of energy to another is called [_____].

Heat energy is a form of energy that is produced by the flow of energy from a warm object to a cooler object. Energy will flow from the warmer object to the cooler object until both objects reach the same [_____]. The transfer, or flow, of energy is caused by the temperature difference between the two objects. Temperature is the measure of how hot or cold a substance is.

Heat also travels between different types of substances. A material that prevents heat from being transferred is called an [_____]. Metal is a poor insulator because it is a conductor which holds heat easily.

Light is another type of energy. Light energy, or radiant energy, is the only type of energy we can see with our eyes. Light energy allows us to fill dark spaces with light, so we can see what is around us.

Light travels in a straight line from its source until it hits an object or surface it cannot penetrate, or pass through. Any material that doesn't allow light to pass through, like concrete, is called [_____].

Sometimes light can hit an object and bounce off its surface instead of passing through it. This phenomenon is called [_____]. Reflection causes light to change direction or bend.

> insulator / opaque / energy conversion / reflection / temperature

⊞ Complete the exercise.

Math Mission

(1) There is a pot of water on the stove. The pot has been heated to 100 degrees F. The water inside the pot is currently 75 degrees F. How many degrees F does the water need to rise so that it is the same temperature as the pot?

Ans. [_____] degrees

(2) There is a pot of water on the stove. The pot has been heated to 80 degrees F. How many degrees F does the water need to cool so that it is room temperature or 65 degrees F?

Ans. [_____] degrees

(3) If the water in the pot was 60 degrees F, how many degrees would it have to rise to be 108 degrees F?

Ans. [_____] degrees

Chapter 4
Forms of Energy

Read the mission. Then, answer the following questions to help you with your solution.

The Mission

You want to build yourself a house that saves money on electricity costs. Use what you learned in this chapter to create a house that uses natural light from the sun to warm and light up the inside. Keep in mind, your house might get very hot during the day, so you'll also want to have a way to reduce the amount of natural light and heat from the sun entering your house.

Before you design...THINK!

1. Describe the mission in your own words.

2. Brainstorm your solution. Write your notes in the space below.

Use the following questions to guide your thinking:

(1) What type of material would you build your house out of? Should it conduct heat or let heat out?

(2) How can you direct more sunlight into your house to help make it brighter?

Chapter 4
Forms of Energy

💡 Read the mission. Then, draw and evaluate your solution.

The Mission

You want to build yourself a house that saves money on electricity costs. Use what you learned in this chapter to create a house that uses natural light from the sun to warm and light up the inside. Keep in mind, your house might get very hot during the day, so you'll also want to have a way to reduce the amount of natural light and heat from the sun entering your house.

Design   Draw or write about your solution below.

Evaluate

What knowledge did you use to design your house? Could you have used a better insulator? How could you have brought more light into the room/house?

Chapter 5
Electricity

Think!

Where does electricity come from?

Check

📖 Read the key points. When you finish, check the box.

Key Points: What is Electricity?

Electricity is a type of energy that can make objects move and work. It is used to provide power to lights and all kinds of machines. Like most forms of energy, we cannot see electricity with our eyes, but we know it is there because of what it does. Electricity exists in two forms: **static** electricity that doesn't move and stays in one place until it can be discharged; and **current** electricity, which moves along a path, called a **circuit**.

In order for current electricity to travel through a circuit it needs a **conductor**. A conductor is a wire that holds an electrical current. Conductors are typically made of metal like copper or aluminum. Most metals conduct electricity, but materials like wood, glass, or plastic do not conduct electricity.

So how do we generate a current for the conductor to carry? One way to generate a current is by using **batteries**. You probably know what a battery is and how it is used. You may use batteries to power your video game controller or your calculator for school. But, do you know how a battery works?

A battery is a container that is made up of multiple cells, or compartments, which store chemical energy that can be converted to electricity. When a battery is plugged into a device, the chemical energy inside it is converted to electrical energy as it runs through the circuit. This powers your calculator or turns on your light bulb. If you remove the battery, you break the circuit and the energy is no longer transferred to your device.

✏️ Complete the exercise.

Test your knowledge

(1) What are the two forms of electricity?

    A. static and conductor   B. static and current

    C. current and conductor   D. current and energy   Ans.

(2) Which form of electricity stays in one place?

    A. static   B. current   Ans.

(3) Choose one of the following that conducts electricity.

    A. glass   B. plastic   C. metal   D. wood   Ans.

(4) Which is an example of a completed circuit?

    A.   B.   C.

Ans.

Chapter 5
Electricity

Think!

Electricity can light up our light bulbs, but what other uses do we have for electricity?

Check

📖 Read the key points. When you finish, check the box.

Key Points: Circuits

Circuits are used to power electrical objects, like lights and computers. As long as the battery is plugged in, the current will continue to move through the circuit and power the object.

However, if you introduce a device called a **switch** into the circuit, you can change or stop the flow of electricity. There are many different ways to use switches in a circuit.

For example, let's say you have multiple light bulbs connected to the same circuit. One option is to place all the light bulbs on the same circuit. We call this a **series circuit**. But this isn't the best option as increasing the number of light bulbs connected in series will dim their brightness since they are all using the same power source. Also, if one bulb breaks, the circuit will be disrupted and none of the other bulbs will light up.

Another option is to place the bulbs on separate circuits. This type of circuit is called a **parallel circuit**. Even if more light bulbs are connected in parallel, their brightness is not affected by the other bulbs. However, this type of circuit drains a battery quicker than a series circuit. But one good thing is that if one light bulb breaks, the other light bulbs will not go out. This is because the current flows through other connected circuits and not through a single circuit that passes through each light bulb.

Engineers use circuit maps to help them figure out how many batteries or what type of circuit to use to power a device. Look at the symbols below to see how engineers design circuits.

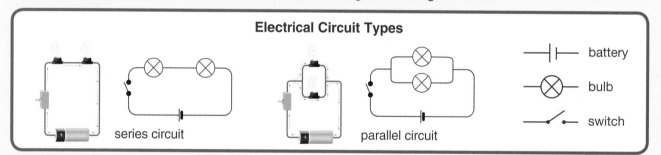

Electrical Circuit Types

series circuit   parallel circuit

battery
bulb
switch

✏️ Complete the exercise.

Test your knowledge

Answer T for true or F for false.

(1) A circuit like Circuit A is called a parallel circuit.

Ans.

(2) In Circuit A, when bulb 1 is burned out, then bulb 2 does not light.

Ans.

(3) In Circuit B, the brightness of the bulbs I and II will be the same.

Ans.

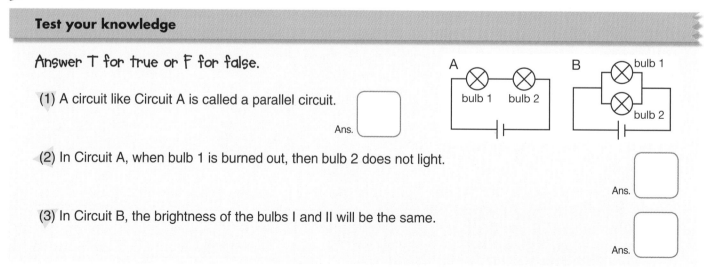

A   bulb 1   bulb 2

B   bulb 1   bulb 2

Chapter 5
Electricity

Think! Can you think of objects around your home that are powered by electricity?

📖 **Read the key points. When you finish, check the box.**

Check ✓

Key Points: Circuits and Power

Remember that electricity is a type of energy that travels along a path, called a circuit. When a battery and a light bulb are correctly connected with a wire conductor, the light bulb will light up and brighten the surroundings.

Circuits can be used to power many different types of devices. If you take a circuit that was used to power a light bulb and connect it to an electronic music box, energy will flow through the circuit and the music box motor will create sound. In this way, electrical energy can be converted into other forms of energy (energy conversion).

Forms of energy include light energy, kinetic energy (motion energy), and sound energy. Electricity can also be converted into heat or thermal energy. For example, a hair dryer generates hot air when it is provided with electricity.

Electricity can also be produced by machines. Making electricity is called **power generation**. Most of the electricity we use in our lives is generated in power plants. At a power plant, electricity is produced by turning the shaft of a generator and converting mechanical energy into electrical energy. Electricity can also be forced into a battery through a process called **charging**. A device that collects and stores electricity is called a capacitor.

power plants

capacitor

✏️ **Complete the exercise.**

Test your knowledge

Choose all of the following types of energy that can be converted from electrical energy.

A. light   B. heat   C. sound   D. motion

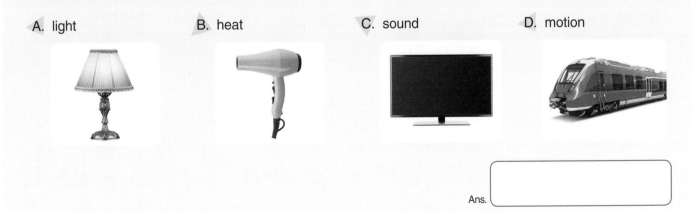

Ans. _____

Electricity

✏️ **Use the word box below to fill in the blanks and review key vocabulary.**

Review the Key Points

Electricity is a type of energy that can make objects move and work. Electricity exists in two forms: static electricity that doesn't move and stays in one place until it can be discharged; and current electricity, which moves along a path, called a _____.

In order for current electricity to travel through a circuit it needs a _____. A conductor is a wire that holds an electrical current. Conductors are typically made of metal like copper or aluminum.

One way to generate a current is by using _____. As long as the battery is plugged in, the current will continue to move through the circuit and power the object. However, if you introduce a device called a switch into the circuit, you can change or stop the flow of electricity.

One option is to place all the light bulbs on the same circuit. We call this a _____ circuit. But this isn't the best option as increasing the number of light bulbs connected in series will dim their brightness since they are all using the same power source. Also, if one bulb breaks, the circuit will be disrupted and none of the other bulbs will light up.

Another option is to place the bulbs on separate circuits. This type of circuit is called a _____ circuit. Even if more light bulbs are connected in parallel, their brightness is not affected by the other bulbs. However, this type of circuit drains a battery quicker than a series circuit. But one good thing is that if one light bulb breaks, the other light bulbs will not go out.

Electrical energy can be converted into other forms of energy (energy conversion).

Electricity can also be produced by machines. Making electricity is called power generation. Electricity can also be forced into a battery through a process called _____. A device that collects and stores electricity is called a capacitor.

> parallel / circuit / series / conductor / charging / batteries

🔢 Complete the exercise.

Math Mission

The chart on the right shows the usable time for each battery.

How many hours does a battery last when used continuously?

Battery	Time
A	7 hours
B	40 hours
C	18 hours
D	2 hours

(1) How many more hours can battery C be used than battery A?

Ans. _____ hours

(2) What is the difference in usable time between the battery with the longest usable time and the battery with the shortest usable time?

Ans. _____ hours

Chapter 5
Electricity

💡 **Read the mission. Then, answer the following questions to help you with your solution.**

The Mission

All of the lights in your house are on a series circuit. One day a light bulb breaks and all the lights go out! Design a better way to run the lights in your house.

Before you design...THINK!

1. Describe the mission in your own words.

2. Brainstorm your solution. Write your notes in the space below.

   Use the following questions to guide your thinking:

   (1) Do you need to change the type of circuit?
   (2) Can you change the type of conductive wire you are using?
   (3) Can you add switches or batteries to help balance the flow of electricity?

Electricity

💡 Read the mission. Then, draw and evaluate your solution.

The Mission

All of the lights in your house are on a series circuit. One day a light bulb breaks and all the lights go out! Design a better way to run the lights in your house.

Design

Draw or write about your solution below.

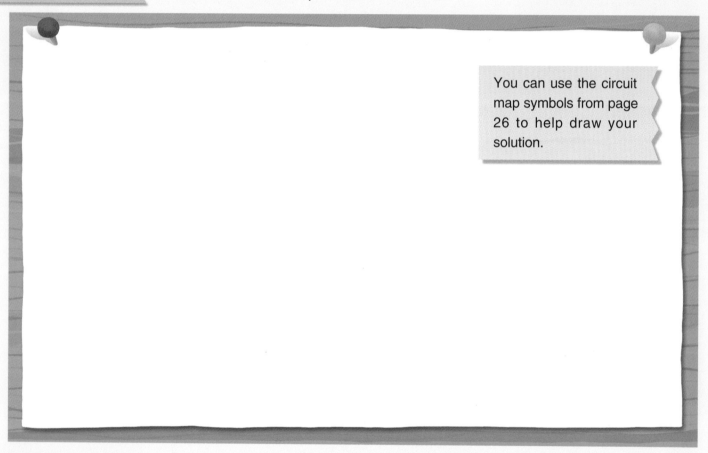

You can use the circuit map symbols from page 26 to help draw your solution.

Evaluate

Did you use the circuit map symbols to help plan your solution? How could you improve your circuit design? Would you add more light bulbs or batteries?

Chapter 6
Sound Waves

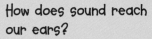
Think!

How does sound reach our ears?

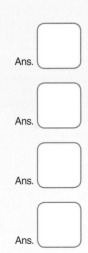
Check

📖 Read the key points. When you finish, check the box.

Key Points: What are sound waves?

We hear various sounds in everyday life. For example, some people may wake up in the morning to the sound of an alarm clock. We hear the voices of our teachers and friends when they speak in school. We use our ears to capture sound.

Sound is produced when something **vibrates** or moves back and forth through a substance like air or water. Think about how a guitar produces sound. When the player's fingers pluck a string, a vibration is generated and you hear a sound. Another example of sound produced by vibration is your voice. We produce sound from our mouths thanks to a vibrating membrane in our throats called a vocal cord. You can even feel it vibrate if you place your fingers on your throat and hum a song!

These vibrations generate sound energy in the form of **sound waves**. Sound energy cannot be seen with our eyes, but it is audible, which means it can be heard. However, sound waves can only be heard when they travel through air, water, or other substances. Sound waves cannot travel to our ears if there is no matter for them to travel through.

✏️ Complete the exercise.

Test your knowledge

Answer T for true or F for false.

(1) A person's voice is one example of sound.

Ans. ☐

(2) Sound is produced by back and forth movement called vibration.

Ans. ☐

(3) Sound waves are waves that can be seen directly by one's own eyes.

Ans. ☐

(4) Sound waves can travel if there is nothing for them to interact with.

Ans. ☐

Sound Waves

Think!

Why are some sounds louder than others? Why are some sounds quieter?

Check

📖 Read the key points. When you finish, check the box.

Key Points: Characteristics of Sound

We can measure various sounds and identify the characteristics of those sounds. Here you will learn some characteristics of sound: "pitch" and "volume".

One of the characteristics of sound is **pitch**. The faster the vibration of the object, the higher pitched the sound will be. On the previous page, you learned that sound waves are generated by an object's vibrations.

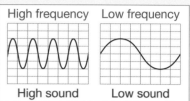

The number of vibrations for each sound pitch is called **frequency**. A higher frequency value is recognized as a higher pitch. High-pitched sounds have faster frequencies or more vibrations than low-pitched sounds with slow frequencies. A whistle is an example of a high-pitch sound and thunder during a storm is an example of a low-pitch sound.

Another characteristic of sound is **volume**. Volume is when a sound gets louder or softer. As you know, sound is created by vibrations and those vibrations affect the volume of a sound. The greater the vibration, the louder the volume. The smaller the vibration, the lower or softer the volume of the sound.

Sounds can also be amplified, or made louder, in several ways. For example, beating a drum with greater force and speed, blowing harder on a recorder, or using more energy when you shout, can all change the loudness or volume of sound.

✏️ Complete the exercise.

Test your knowledge

(1) Which wave has the highest frequency? Choose the letter of the best answer.

A.

B.

C.

Ans.

(2) Which wave has the loudest sound? Choose the letter of the best answer.

A.

B.

C.

Ans.

(3) Choose one wrong way to increase the volume of sound.

   A. Beating the drums with great force.

   B. Blowing the recorder softly.

   C. Shouting with more energy.

Ans.

Chapter 6
Sound Waves

Think!

Are there other ways to change the volume of sounds?

Check

📖 Read the key points. When you finish, check the box.

Key Points: More Characteristics of Sound

Remember that sound waves have to travel through a substance to your ear. And they can travel through different kinds of substances, including things that are solid like metal, liquid like water, or gaseous like air.

When a sound wave hits the surface of an object, some of its energy bounces off the surface. This is called sound **reflection**. If a hard object such as a ball or a stone hits a solid surface like metal or concrete, you will hear a sharp and loud sound.

On the other hand, if the ball hits something soft such as sand or grass, you will hear less of the sound. This is because the energy of the sound wave enters the surface of the object and spreads out. This is called sound **absorption**.

The amount of sound wave energy that is reflected or absorbed depends on the material on the surface of the object. When hitting a hard, smooth material such as metal, much of the energy of the sound wave is reflected.

Another characteristic of sound is **resonance**. Resonance is caused by sound waves that repeatedly hit multiple hard, smooth surfaces at the same time. Have you ever listened to your singing voice reverberating off the walls when you sing in the shower? This is an example of resonance.

A final characteristic of sound is distortion. When you are underwater and you hear someone yelling at you from above the water their voice may sound weird. This is because water disrupts sounds that travel through it. When sound is disrupted as it travels through certain materials, like water, it is called **distortion**.

✏️ Complete the exercise.

Test your knowledge

Identify the property of sound based on the descriptions below.

> A. reflection   B. absorption   C. resonance   D. distortion

(1) When a sound wave hits the surface of an object and much of its energy enters the surface and is muffled.

Ans. ☐

(2) When a sound is disrupted as it travels through matter like water.

Ans. ☐

(3) It is created by a sound wave repeatedly hitting the surface of the object.

Ans. ☐

(4) When a sound wave hits the surface of an object and much of its energy bounces off the surface.

Ans. ☐

Chapter 6
Sound Waves

🖉 **Use the word box below to fill in the blanks and review key vocabulary.**

Review the Key Points

Sound is produced when something [＿＿＿] or moves back and forth through a substance like air or water.

These vibrations generate sound energy in the form of [＿＿＿＿]. Sound energy cannot be seen with our eyes, but it is audible, which means it can be heard. However, sound waves can only be heard when they travel through air, water, or other substances. Sound waves cannot travel to our ears if there is no matter for them to travel through.

One of the characteristics of sound is [＿＿＿]. The faster the vibration of the object, the higher pitched the sound will be.

The number of vibrations for each sound pitch is called [＿＿＿＿]. A higher frequency value is recognized as a higher pitch. High-pitched sounds have faster frequencies or more vibrations than low-pitched sounds with slow frequencies.

Another characteristic of sound is [＿＿＿＿]. Volume is when a sound gets louder or softer. The greater the vibration, the louder the volume. The smaller the vibration, the lower or softer the volume of the sound.

The amount of sound wave energy that is reflected or absorbed depends on the material on the surface of the object. When hitting a hard, smooth material such as metal, much of the energy of the sound wave is reflected.

Another characteristic of sound is [＿＿＿＿]. Resonance is caused by sound waves that repeatedly hit multiple hard, smooth surfaces at the same time.

> volume / vibrates / resonance / frequency / pitch / sound waves

🖩 **Complete the exercise.**

Math Mission

Did you know that sound can travel 340 meters per second through the air? Let's see how fast the sound from certain objects can travel! You can refer to the chart on the right for help with multiplication formulas and answers.

(1) Tom heard the sound two seconds after seeing the fireworks. How far is the place where the fireworks occurred from the place where he stands?

$$340 \times 2 = \boxed{}$$

Ans. [＿＿＿] m

(2) Nancy heard the sound 5 seconds after seeing the lighting flash. How far is the place where the lightning flashed from the place where she stands?

[＿＿＿] × [＿＿] = [＿＿＿]

Ans. [＿＿＿] m

Multiplication Chart

$340 \times 1 = 340$
$340 \times 2 = 680$
$340 \times 3 = 1020$
$340 \times 4 = 1360$
$340 \times 5 = 1700$

Sound Waves

Read the mission. Then, answer the following questions to help you with your solution.

The Mission

Design an original instrument that can produce various sounds. You want to enjoy a variety of sounds with a single instrument, such as high, low, loud, soft, reverberating, and quickly fading.

Before you design...THINK!

1. Describe the mission in your own words.

2. Brainstorm your solution. Write your notes in the space below.

   Use the following questions to guide your thinking:

   (1) What type of material can you use to make your instrument louder? Softer?
   (2) How can you amplify the sound from your instrument?

Chapter 6

Sound Waves

💡 Read the mission. Then, draw and evaluate your solution.

The Mission

Design an original instrument that can produce various sounds. You want to enjoy a variety of sounds with a single instrument, such as high, low, loud, soft, reverberating, and quickly fading.

Design

Draw or write about your solution below.

Evaluate

Do you think your instrument was successful in creating multiple types of sound? What part of your instrument could you change to make it work better?

Chapter 7
Structures of Matter

Think!

Have you ever thought about what objects are made of?

📖 **Read the key points. When you finish, check the box.**

Check ✓

Key Points: What is matter?

Look around the room you are in. What do you see? If you are in a living room, do you see a couch or chair? A rug and a lamp? If you are in your bedroom, do you see a desk? Your bed or closet? Do you see books or video games on your floor?

Now, did you know all those things you can see are made of matter? And if you did know, do you know what matter is?

Let's look at the things around you once again. Each object occupies space in the room. However, a notebook takes up less space in a bedroom than your bed and a couch take up a lot more space in a living room than a lamp. This is because all objects have **volume**. Volume is the amount of space an object occupies or takes up.

In addition to volume, all matter also has **mass**. Mass is the amount of matter in an object or substance. When an object has a large mass, it feels heavy. The opposite is true for a small object. If the object is small, the mass will be light. For example, a notebook is lighter than a desk.

These two characteristics are used to define matter. So, to answer the question above, **matter** is any physical object or substance that has mass and occupies space (volume).

✏️ **Complete the exercise.**

Test your knowledge

(1) Choose the correct explanation for matter.

A. Volume represents the number of each item.
B. Mass is the amount of space occupied by an object.
C. Every object has a volume and a mass.

Ans. [ ]

(2) Choose all of the following that are made of matter.

A. computer   B. apple   C. bicycle   D. rock

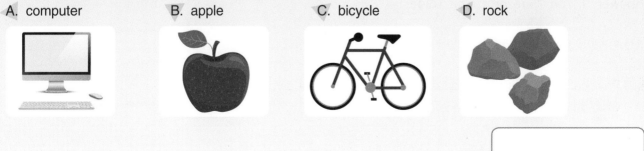

Ans. [ ]

Chapter 7
Structures of Matter

Think!

How can you determine how much mass or volume an object has?

Check

📖 **Read the key points. When you finish, check the box.**

Key Points: Physical Properties

You have learned that an object has volume and mass. But, did you know objects and substances also have other physical properties you can see? **Physical properties** are properties of an object that can be measured or seen without changing the matter it is made up of. Some examples of physical properties are mass, volume, color, smell, and texture.

Physical Properties

See
New car is shiny

Touch
The ceramic pot is hard

Hear
A bell makes noise

Smell
Roses smell sweet

Feel
The cat is soft

Measure
The temperature is low

Physical properties are important because they can help you identify what an object or substance is. Each substance has its own unique set of properties that define it. Let's use the example of the notebook again. What makes your notebook different from your desk?

It is smaller, the pages are white and soft, and the cover is hard. You also know that the notebook is lighter than the desk. These are properties that define the notebook.

Scientists have created many different lists and scales to help us define an object by its physical properties. Let's look at the Mohs Hardness Scale on the right. Scientists use this scale to help determine how hard or soft a rock or mineral is compared to other rocks and minerals.

Mohs Hardness Scale

Name	Scale Number	Common Object
Diamond	10	
Corundum	9	← Masonry Drill Bit
Topaz	8	
Quartz	7	← Steel Nail
Orthoclase	6	← Knife
Apatite	5	
Fluorite	4	← Copper Coin
Calcite	3	← Fingernail
Gypsum	2	
Talc	1	

Increasing Hardness

✏️ **Complete the exercise.**

Test your knowledge

Answer T for true or F for false.

(1) Volume and Mass are example of physical properties.

Ans.

(2) Physical properties are important because they can help you identify what an object or substance is.

Ans.

(3) According to the Mohs Hardness Scale, talc is harder than diamond.

Ans.

Structures of Matter

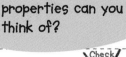

What other physical properties can you think of?

📖 **Read the key points below. When you finish, check the box.** ✓

Key Points: Chemical Properties

Physical properties are not the only properties of matter. All substances also have **chemical properties** that are special to each substance. Chemical properties are the properties of a substance that can be observed when a substance goes through a chemical change. A chemical change is a type of change that also changes the physical properties of a substance.

An example of a chemical change you may have seen before is when metal rusts. Have your ever left your bike out in the rain or snow for a few days? When you go to ride it again you may have seen an orange-red colored material built-up on the bike chain. This is rust. Rust is the result of a chemical reaction between the metal of your bike and precipitation like rain or snow. And once metal rusts you cannot change it back. This is what makes rust a chemical change.

So why do substances have chemical properties? To answer that you must look at what a substance is made up of. Everything around you is made up of tiny particles you can't see. These particles are called **atoms**.

Examples of Chemical Properties

- Ability to rust
- Ability to decompose
- Reacts with water, oxygen, acids, bases, or other substances
- Flammability

Atoms are the building blocks for substances called **elements**. Elements have specific chemical and physical properties that cannot be broken down into other substances through ordinary chemical reactions. All substances in the world are made of elements. Examples of elements include hydrogen, oxygen, aluminum, calcium, and iron. There are more than 120 elements!

Elements react differently when they encounter other elements. For example, your bike rusts because it is made of iron and the physical properties of iron change when it meets with water and oxygen. These three substances react when they come together to change the physical properties of the bike and form rust.

✏️ **Complete the exercise.**

Test your knowledge

Match the terms below to the correct definition.

> A. chemical properties   B. atoms   C. elements

(1) These have specific chemical and physical properties that cannot be broken down into other substances through ordinary chemical reactions. All substances in the world are made of these.

Ans. ☐

(2) These are made up of tiny particles you can't see. These are the building blocks for substances called elements.

Ans. ☐

(3) These are the properties of a substance that can be observed when a substance goes through a chemical change.

Ans. ☐

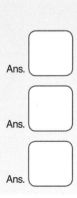

Structures of Matter

✏️ Use the word box below to fill in the blanks and review key vocabulary.

Review the Key Points

All objects have [   ]. Volume is the amount of space an object occupies or takes up.

In addition to volume, all matter also has [   ]. Mass is the amount of matter in an object or substance.

[   ] properties are properties of an object that can be measured or seen without changing the matter it is made up of. Physical properties are important because they can help you identify what an object or substance is. Each substance has its own unique set of properties that define it.

Scientists have created many different lists and scales to help us define an object by its physical properties.

All substances also have [   ] properties that are special to each substance. Chemical properties are the properties of a substance that can be observed when a substance goes through a chemical change. A chemical change is a type of change that also changes the physical properties of a substance.

Everything around you is made up of tiny particles you can't see. These particles are called [   ].

Atoms are the building blocks for substances called [   ]. Elements have specific chemical and physical properties that cannot be broken down into other substances through ordinary chemical reactions. All substances in the world are made of elements. Elements react differently when they encounter other elements.

mass / atoms / volume / elements / chemical / physical

 Complete the exercise.

Math Mission

Scientists can measure different properties of objects. One physical property they often measure is weight. Read the scales below and record the weight.

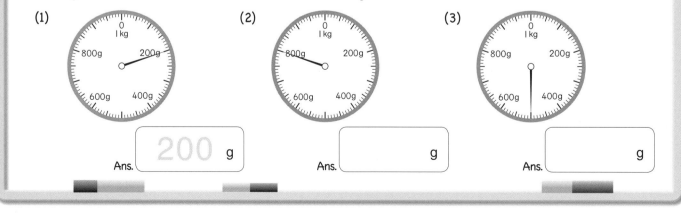

(1)

200 g

Ans.

(2)

g

Ans.

(3)

g

Ans.

Structures of Matter

💡 Read the mission. Then, answer the following questions to help you with your solution.

The Mission

As a scientist, you have successfully produced a new substance in the lab! Use what you have learned about the properties of matter to describe your substance and define its characteristics, so others will be able to identify it in the future.

Before you design...THINK!

1. Describe the mission in your own words.

2. Brainstorm your solution. Write your notes in the space below.

Use the following questions to guide your thinking:

(1)   What will you name your substance?

(2)   What color is it? What is its mass? What is its volume? Is it magnetic?

(3)   What chemical properties could it have? Does it dissolve in water? Or rust in the rain?

Structures of Matter

-ON- Read the mission. Then, draw and evaluate your solution.

The Mission

As a scientist, you have successfully produced a new substance in the lab! Use what you have learned about the properties of matter to describe your substance and define its characteristics so others will be able to identify it in the future.

Design

Draw or write about your solution below.

Evaluate

What substance did you create? Did you give your substance physical and chemical properties? How would your substance react with other substances?

Chapter 8
Changes in States of Matter

Think!

Can you name the three main states of matter?

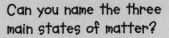

📖 Read the key points. When you finish, check the box.

Check ✓

Key Points: States of Matter

Most substances on Earth can be classified as either a **solid**, a **liquid**, or a **gas**. These are the three primary states of matter. Let's look at the characteristics of solids, liquids, and gases.

SOLID   LIQUID   GAS

A **solid** is a substance with a defined shape and volume. The atoms in a solid stick together in a specific pattern and do not move very much. Solids don't change shape unless great force or heat is applied to them. Examples of solids are ice, wood, bricks, apples, notebooks and many other objects.

A **liquid** is a substance that has a defined volume but no defined shape. The atoms in a liquid are less tightly packed than in a solid, but still close together. They can move a little bit because they have more space between their atoms than solids. That is why liquids can change shape. For example, when you pour juice into different size glasses, the juice will fit the shape of each glass. Some examples of liquids are water, juice, soda, coffee, and tea.

A **gas** is a substance that does not have a defined shape or volume. The atoms of a gas are spread apart from each other and do not stick together like in a solid. They do not have a fixed shape because they move around, and the pattern and volume of the arrangement of each atom is not concrete. Some examples of gases are air, steam, oxygen, and carbon dioxide.

✏️ Complete the exercise.

Test your knowledge

Match the states of matter to each example.

(1) solid ●

● milk

(2) liquid ●

● watch

(3) gas ●

● air

Chapter 8
Changes in States of Matter

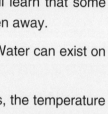

Think!

How does matter change form?

Check

📖 **Read the key points. When you finish, check the box.**

Key Points: Properties of States of Matter

Remember there are three main states of matter: solid, liquid, and gas. Here you will learn that some substances can change from one state of matter to another when energy is added or taken away.

The best substance to use when looking at how states of matter can change is water. Water can exist on Earth in all three states of matter as a solid, a liquid, and a gas.

You know that water is liquid. What happens if you heat the water? As the water warms, the temperature rises. When the temperature of the water reaches around 100 degrees Celsius (212 degrees Fahrenheit), bubbles are produced from the water. This action is called **boiling**. The bubbles that rise to the surface when the water boils are water in the form of a gas called water vapor. The process of water turning into water vapor is called **evaporation**. When water is heated, it evaporates and transforms from liquid to gas.

Water can also be cooled. Cooling water will lower its temperature. When the temperature drops to 0 degrees Celsius (32 degrees Fahrenheit), the water turns into ice. You have already learned that ice is a solid. When water is cooled, it changes its form from liquid to solid.

So, water can be a solid (ice), a liquid (water), and a gas (water vapor) depending on the temperature. In addition to water, there are other substances that change form depending on the temperature. For example, the solid wax of a candle melts from the heat of its flame and becomes a liquid wax.

Gas

Liquid

Solid

✏️ **Complete the exercise.**

Test your knowledge

Choose the best word to complete each sentence.

(1) If water is kept heated, it will boil at about (0 / 100) degrees Celsius.

Ans.

(2) Liquid water turns into a (solid / gas) called water vapor as it continues to be heated.

Ans.

(3) Liquid water turns into (solid / gas) ice as it cools.

Ans.

(4) The state of matter of water can change from a solid, a liquid or a gas depending on (temperature / amount).

Ans.

Changes in States of Matter

Think!

When a substance changes from one form of matter to another, do its properties also change?

Check

📖 **Read the key points. When you finish, check the box.**

Key Points: Properties of Matter Continued

Remember that all matter has a volume and a mass. You also learned that the state of matter (solid, liquid, gas) can change as the temperature changes.

When a substance changes states, it also changes volume. Substances often increase in volume when they are warmed and decrease when cooled. For example, if a container full of air is warmed, the container will expand because the atoms in the air move faster. The opposite happens if you cool a container full of air. The atoms that make up the air will slow down and the container may collapse in on itself a little. This means that the volume increases when the gas is warmed and decreases when the air is cooled. In this case, we can see the change in the volume of air with our eyes.

If we only use the example of air, you can say that the volume of a substance increases as it changes states. However, water does not follow this rule. When water changes from a solid to a liquid, its volume decreases. In other words, ice has a larger volume than liquid water. Want to see this idea in action for yourself? Try filling a plastic cup with water and put it into the freezer. As it freezes you can see how the ice swells and rises in the glass. When water changes from a liquid to a solid, the volume increases.

Even as the state of matter changes, the mass does not change. Whether the water is warmed or cooled, the number of atoms in it does not increase or decrease. The atoms simply speed up or slow down. So, the mass of a substance always stays the same.

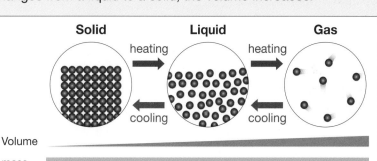

✏️ **Complete the exercise.**

Test your knowledge

Choose the word that best complete the sentence.

(1) When a gas is warmed, the volume will (increase / decrease / not change).

Ans.

(2) When solid ice changes into liquid water, the volume will (increase / decrease / not change).

Ans.

(3) When water is cooled, the mass will (increase / decrease / not change).

Ans.

Changes in States of Matter

✏️ **Use the word box below to fill in the blanks and review key vocabulary.**

Review the Key Points

Most substances on Earth can be classified as either a solid, a liquid, or a gas.

A [_____] is a substance with a defined shape and volume. The atoms in a solid stick together in a specific pattern and do not move very much.

A [_____] is a substance that has a defined volume but no defined shape. The atoms in a liquid are less tightly packed than in a solid, but still close together. They can move a little bit because they have more space between their atoms than solids.

A [_____] is a substance that does not have a defined shape or volume. The atoms of a gas are spread apart from each other and do not stick together like in a solid. They do not have a fixed shape because they move around, and the pattern and volume of the arrangement of each atom is not concrete.

When water is heated, it [_____] and transforms from liquid to gas. When water is cooled, it changes its form from liquid to solid.

Water can change into solid (ice), liquid (water), and gas (water vapor) depending on the temperature. In addition to water, there are other substances that change form depending on the temperature.
As the substance changes state, the volume changes. Substances increase in volume when warmed and decrease when cooled.

When a substance changes states, it also changes [_____]. Substances often increase in volume when they are warmed and decrease when cooled. However, water does not follow this rule. When water changes from a solid to a liquid, its volume decreases.

Even as the state of matter changes, the [_____] does not change. Whether the water is warmed or cooled, the number of atoms in it does not increase or decrease.

> gas / mass / liquid / evaporates / solid / volume

🖩 **Complete the exercise.**

Math Mission

Look at the table with the results of a study of the length of time water was heated and the temperature of the water. Which is correct when graphing this result?

heating time (minute)	0 min.	3 min.	6 min.	9 min.	12 min.	15 min.	18 min.	21 min.	24 min.	27 min.	30 min.
water temperature (°C)	19°C	35°C	58°C	72°C	85°C	91°C	98°C	99°C	99°C	99°C	99°C

Ans. [_____]

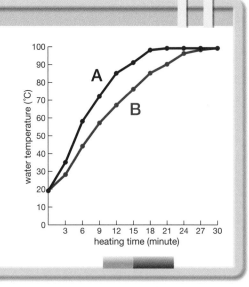

Chapter 8

Changes in States of Matter

Read the mission. Then, answer the following questions to help you with your solution.

The Mission

You are in charge of feeding your baby brother lunch. He's crying for a jam sandwich and won't stop crying until he gets it! But the jam jar is stuck shut! Think of a way to open the sticky jam jar using what you have learned in Chapter 8 to help you solve this problem.

Before you design...THINK!

1. Describe the mission in your own words.

2. Brainstorm your solution. Write your notes in the space below.

   Use the following questions to guide your thinking:

   (1) How can you change the properties of the jar to help open the jar?
   (2) Can you heat or cool the jar?
   (3) What forms of matter are you dealing with? Solids, liquids, or gases? Can you change the properties of one of them to help you?

Chapter 8
Changes in States of Matter

💡 Read the mission. Then, draw and evaluate your solution.

The Mission

You are in charge of feeding your baby brother lunch. He's crying for a jam sandwich and won't stop crying until he gets it! But the jam jar is stuck shut! Think of a way to open the sticky jam jar using what you have learned in Chapter 8 to help you solve this problem.

Design

Draw or write about your solution below.

Evaluate

How did you open the jar? Did you heat it to loosen the lid? Or freeze it so the jar would expand? Do you think your plan would cause the jar to break?

Chapter 1 Forces and Motion

1

Test your knowledge

(1) B   (2) A

2

Test your knowledge

(1) balanced   (2) balanced   (3) unbalanced

3

Test your knowledge

(1) B   (2) A

4

Review the Key Points

force / Direction / balanced / unbalanced / Friction

Math Mission

(1) $8\frac{1}{2}$ cm (8.5 cm)   (3) 3 cm

(2) 5 cm   (4) tile floor

5 (Sample Response)

Before you design... THINK!

1. Use knowledge of forces to build the fastest sled and win the race.

2. I could make my sled legs out of metal so it glides on the icy parts and the top part out of wood so it weighs less.

6 (Sample Response)

Design

• My sled legs would be made of metal like an ice skate blade to go fast on the icy parts of the hill.

• I would build the top part of my sled out of wood so it was lighter. This can help it go faster too.

Evaluate

My sled would win the race! If I had to change part of my design, I would find a way to make my sled more slippery so it creates less friction.

Chapter 2 Simple Machines

7

Test your knowledge

(1) force   (2) done   (3) not done

8

Test your knowledge

(1) B   (2) C   (3) A

9

Test your knowledge

(1) inclined plane

(2) wedge

(3) screw

10

Review the Key Points

Work / simple / lever / wedge / compound

Math Mission

(1) $10 \div 2 = 5$ $30 \div 5 = 6$ Ans. 6 pounds

(2) $8 \div 4 = 2$ $40 \div 2 = 20$ Ans. 20 pounds

11 (Sample Response)

Before you design... THINK!

1. Find a way to open the chest you found with the help of simple machines.

2. I could use a lever and a pulley to open the chest. I don't think a screw would be helpful.

12 (Sample Response)

Design

Use a wedge to try the chest open, then use a pulley to help lift the heavy lid.

Evaluate

I used a wedge and a pulley to open the chest. I could've used a lever to help open the chest. It could've made it easier.

Chapter 3 **Magnets**

13
Test your knowledge
B, D, E

14
Test your knowledge
(1) repel

(2) attract

(3) attract

(4) repel

15
Test your knowledge
A, B, C, D, E, F, G, H (all)

16
Review the Key Points
magnetic field / metals / north pole / Magnets

Math Mission
(1) Yes   (2) No

17 (Sample Response)
Before you design... THINK!
1. Create a device to help pick up all the pins quickly and without getting hurt.
2. I would use multiple magnets for this plan. By using more magnets, I can pick up more pins at once.

18 (Sample Response)
Design
Glue magnets to the end of a rake or several sticks to move over the rug and collect the pins.

Evaluate
I think my tool would pick up most of the pins easily. I think I would need to add something to the tool to help make sure all the pins are picked up, like the screen on a metal detector.

Chapter 4 **Forms of Energy**

19
Test your knowledge

20
Test your knowledge
A, B, D

21
Test your knowledge
(1) T   (3) F   (5) T

(2) F   (4) T

22
Review the Key Points
energy conversion / temperature / insulator / opaque / reflection

Math Mission
(1) $100 - 75 = 25$ Ans. 25 degrees

(2) $80 - 65 = 15$ Ans. 15 degrees

(3) $108 - 60 = 48$ Ans. 48 degrees

23 (Sample Response)
Before you design... THINK!
1. Create a way to keep your small house warm with energy.
2. I would build my house out of material that conducts heat well so it stays in the room.

24 (Sample Response)
Design
- Metal is a good conductor, so I would build the outside of the house out of metal to help conduct heat and keep the house warm.
- I would put big windows in the walls and roof to help bring in more light for the metal walls to absorb.

Evaluate
I would need to use specific conductors and insulators to make my solution work. I'm not sure how I would light the room when it is not sunny.

Chapter 5 **Electricity**

25

Test your knowledge

(1) B   (2) A   (3) C   (4) C

26

Test your knowledge

(1) F   (2) T   (3) T

27

Test your knowledge

A, B, C, D (all)

28

Review the Key Points

circuit / conductor / batteries / series / parallel / charging

Math Mission

(1) 18 − 7 = 11   Ans. 11 hours

(2) 40 − 2 = 38   Ans. 38 hours

29 (Sample Response)

Before you design... THINK!

1. Create a plan to fix the broken light circuit in the house.
2. I can use parallel circuits and more batteries to help fix the light issues.

30 (Sample Response)

Design

Use the symbols from the circuit map to plan a better circuit layout for the light bulbs in the house. Make sure there are enough batteries to support all the lights.

Evaluate

I think my plan will help the lights work better in the house.

Chapter 6 **Sound Waves**

31

Test your knowledge

(1) T   (2) T   (3) F   (3) F

32

Test your knowledge

(1) C   (2) C   (3) B

33

Test your knowledge

(1) B   (2) D   (3) C   (4) A

34

Review the Key Points

vibrates / sound waves / pitch / frequency / volume / resonance

Math Mission

(1) 340 × 2 = 680   Ans. 680 m

(2) 340 × 5 = 1700   Ans. 1700 m

35 (Sample Response)

Before you design... THINK!

1. Create an original instrument that can make lots of types of sounds.
2. I could use hard materials to make sounds louder or soft materials to make lower sounds.

36 (Sample Response)

Design

• My instrument is like a drum, but you can take the center piece out to make the sound louder or softer. The piece is made of cloth that will make the sound softer.
• If the piece is remove it allows for the sound to echo and be louder.

Evaluate

I think my instrument was successful in making different types of sounds. I could improve my instrument by creating more ways to make it sound louder.

Chapter 7 **Structures of Matter**

37
Test your knowledge
(1) C   (2) A, B, C, D (all)

38
Test your knowledge
(1) T   (2) T   (3) F

39
Test your knowledge
(1) C   (2) B   (3) A

40
Review the Key Points
volume / mass / Physical / chemical / atoms / elements

Math Mission
(1) 200 g   (2) 800 g   (3) 500 g

41 (Sample Response)
Before you design... THINK!
1. Define the properties of the new substance you created.
2. I will give my substance physical and chemical properties. It will have a color, a hardness, and a chemical reaction to rain.

42 (Sample Response)
Design
My substance is called floriandizole. It is purplish in color and harder than diamonds. But it has a chemical property, it can be dissolved in water.

Evaluate
I think my substance would react well with other substances, but it could not be used to build anything outside because rain would melt it.

Chapter 8 **Changes in States of Matter**

43
Test your knowledge
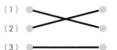
(1)
(2)
(3)

44
Test your knowledge
(1) 100   (3) solid
(2) gas   (4) temperature

45
Test your knowledge
(1) increase   (2) increase   (3) not change

46
Review the Key Points
solid / liquid / gas / evaporates / volume / mass

Math Mission
A

47 (Sample Response)
Before you design... THINK!
1. Find a way to open the stuck jam jar using properties of matter.
2. The jar is stuck closed, but the jam is kind of a liquid. Maybe I could heat or cool it to help open it?
Would I need other tools? Like a pot?

48 (Sample Response)
Design
I would put the jam jar in a pot of water and boil it. I would hope that the solid sticky jam on the lid warms up and changes to a liquid jam. This would allow me to open the jar!

Evaluate
I think changing the properties of the jam by heating it would allow me to open it and solve the problem. I think I could have also put the jar in the freezer to see if the jam would expand as it froze and pop the lid off.